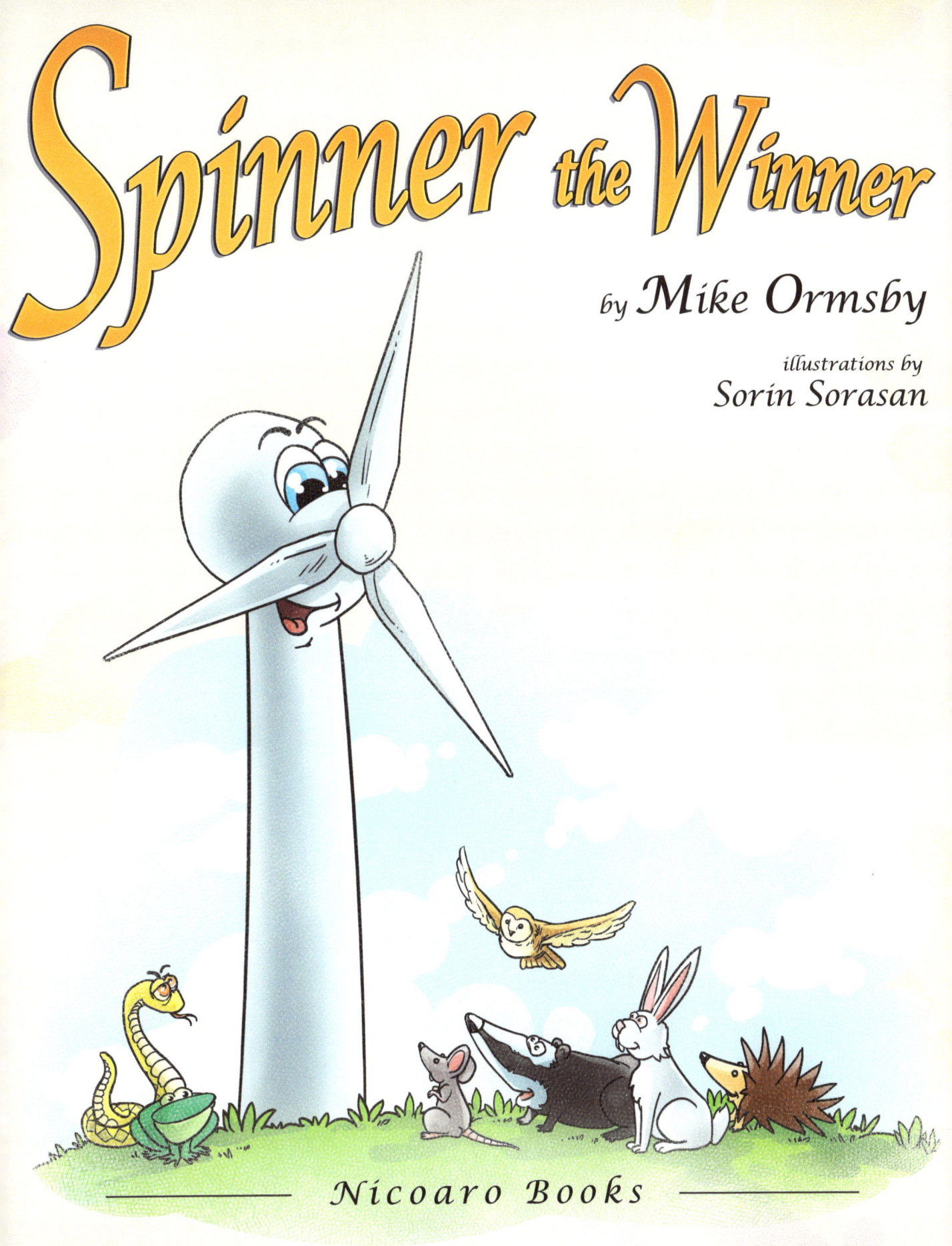

Acknowledgments

Thank you to Judy McCarthy in Boston and to our young readers, Luca & Lara in Bucharest, Dan in London, and Ilinca, Mark & Steven in Langley, for kind words and helpful hints. We are also grateful to Julie Lewthwaite, for proofreading and for helping to format the text.

©Mike Ormsby, 2012
ISBN-13: 978-1478284260

www.mikeormsby.net
spinnwinn@gmail.com
Twitter: @OrmsbyMike

Cover & interior design by Sorin Sorasan
www.sorinsorasan.com

Contents

1. A Home on the Beach! 6
2. A Big Surprise 8
3. A New Home 12
4. New Friends 14
5. Trouble in Biffletop 18
6. Bigger Squeaks 20
7. Smile, Please! 22
8. Red Paint and Rude Words 24
9. No Time for Munch-time! 26
10. Crack! Bang! Hiss! 30
11. Darkness Falls 34
12. People Power 38
13. Yum-Yum, Oil in My Tum! 42
14. A Lucky Escape 46
15. Spinner the Winner! 50
16. Back to the Beach 56

1. A Home on the Beach!

Once upon a windy day, Spinner was riding with his family in a long red truck. It was the first time he had been outside the turbine factory, so he was feeling quite excited, but a little puzzled.

"Where are we going?" he said.

"To live at the beach!" replied his mom. "And when the wind blows, we'll turn our arms to make electricity. That's our job, we are wind turbines!"

"What's electricity?" Spinner asked.

"It provides power to make things work," his dad answered. "Fridges and computers, televisions and traffic lights, cinemas and things like that."

"Will I see the ocean?"

"Yes, and a funfair!" said Spinner's sister, Turner.

Joe the driver tooted his horn and waved his spotty green cap and Spinner yelled, "Yippee! This is the best day of my life!"

2. A Big Surprise

After a long trip along a bendy road, the truck stopped on high cliffs above a golden beach, where the blue ocean sparkled like diamonds. Spinner watched a yellow crane lifting big turbines into holes dug especially for them. A man dressed all in black walked up and down. His name was Riley Rules and he was checking boxes on forms, because that was his job and he was very good at it.

But today, Riley was feeling puzzled because something was wrong. He looked at his forms and he looked at the turbines. Finally, he looked at Joe and said in a rather bossy voice, "You brought too many. I don't need that little turbine. A storm will blow it over. Take it to Biffletop Hill."

Joe scratched his head. "Biffletop Hill? That's a long way."

"Use a map," Riley said, "and your brain if you've got one. Off you go, nincompoop."

Soon Joe drove away and Spinner was sitting all alone in the back of the truck, because he was just a little turbine, no use to anyone, especially in a storm. He watched his mom, dad and sister Turner slowly spinning their arms to say goodbye from their new home on the high cliffs.

"Please stop, Joe!" Spinner cried. "I want to live by the beach with Mom and Dad and Turner!"

But Joe did not stop; he drove the red truck far away over the green hills and big wobbly tears were soon trickling from little Spinner's eyes.

3. A New Home

Spinner fell asleep and did not wake up until the truck stopped with a bump that made him jump.

Biffletop Hill was a dark and silent place, until Joe opened the tailgate of his truck — CLANK-CLANK — and an owl hooted from the trees.

A yellow crane lifted Spinner into a hole and Joe said, "It's very late, my friend. I must go home, sorry about the mix-up!" He drove quickly away without giving the little turbine even one drop of oil.

Spinner stood alone under a black sky. Heavy raindrops fell on his head and trickled down his neck and his arms squeaked when they turned.

"This is the worst day of my life!" he sniffled to himself.

The raindrops dripped from his nose and landed on the grass: PLOP! They plopped so hard, Spinner did not hear the owl hooting again across the dark valley to announce: *Something strange has arrived!*

Nor did he notice all the bright little eyes watching from the hedgerows.

4. New Friends

Next morning, warm sunshine dried the raindrops from Spinner's face and he felt a bit better, especially when some animals came to say hello. It was nice to make new friends, although they did ask some strange questions.

"So… are you an airplane?" asked Flash Rabbit, hopping in circles, his white tail bobbing.

"No," replied Spinner.

"Are you a television transmitter?" said Marjorie Mouse, twiddling her whiskers and standing on her hind legs for a better view.

"No," Spinner answered.

"A spaceship?" murmured Prickle Hedgehog and Spinner tried not to laugh, as he replied, "Not quite."

Tuft the Barn Owl turned his head right round and blinked his big yellow eyes and said, "He's a helicopter. Any fool can see that."

"Speak for yourself," said Prickle Hedgehog, sniffing for rocket paint.

Big Badger twitched his snout and said, "I don't want to be too black and white about all this, but let's try an easy question: what's your name?"

"My name is Spinner."

"Spinner the what?" inquired Frank Frog.

"I'm Spinner the Wind Turbine."

The animals looked at each other as if to say *Spinner-the-who?*

"So nice to meet you, Spinner," hissed Snake Adam. "But personally, I cannot help wondering… what's your businesssss?"

The other animals frowned because Snake Adam could be a rude and rather slippery fellow.

"If you mean my job," replied Spinner, with a big smile, "I make electricity from the wind!"

"You make a lot of noise," hissed Snake Adam. Flash Rabbit nodded and said, "I agree. My ears are lop-sided from lack of sleep."

Spinner quickly apologized. "I'm very sorry, but you see Joe Driver forgot my oil. That's why last night, I was squeaking a little."

"You were sssqueaking a lot," said Snake Adam. "I hardly ssslept."

Prickle Hedgehog rolled sideways and whispered, "No need to be rude, Adam."

"Spinner, where are your electric wires?" asked Marjorie Mouse.

"They're underground! It's very clever and modern," Spinner replied proudly. "Some people say I'm the way forward."

"Perhaps you're the way backwardsss," said Snake Adam.

"We need a discussion," said Tuft. "Gather round, everyone."

The animals huddled to talk about *The Way Forward or Perhaps Backwards, Something Strange Anyway,* and while they chattered and nattered, Spinner surveyed his new home. It wasn't so bad, in the warm light of morning. The green hills were nice, but there was not even one friendly turbine waving hello, just a line of wooden poles with nasty looks that seemed to say: *We make the electricity round here, not you.*

"Excuse me," Spinner said to his new friends, "but I have a question, too. What's that place called?"

They looked down the hill at a busy little town with sharp red roofs. "Biffletop, of course," said Tuft the Barn Owl, who knew everything.

But he did not know that some Biffletoppers were unhappy this morning. Can you guess why?

5. Trouble in Biffletop

The Lord Mayor of Biffletop was sitting in his office and feeling rather worried, because he had four visitors, all complaining in loud voices.

His face was pink with embarrassment and he wished he were home watching TV, not here listening to a big fuss about a little turbine.

"Mister Mayor!" yelled Jack Throttle the taxi driver, "why is that stupid-windmill-thing on Biffletop Hill? It makes too much noise!"

"It also spoils our nice view!" howled Betty Baker.

Miss Cane the headmistress poked the Mayor with a long finger and said, "Most of the schoolchildren could not stay awake in class today! They were very sleepy because of *squeaky-squeaky*, all night long!"

"Mister Mayor, what will you do?" demanded Johnny Giblets, a well-known chicken farmer whose pointy beak was perfect for picking up gossip.

The Lord Mayor stood up straight and tucked his thumbs under the shiny gold chain that he liked to wear around his neck, even at bedtime.

"Quiet, you noisy Biffletoppers! I am your Lord Mayor and I deserve more respect!" he said. "I promise I will stop those squeaks, later. But now, I have more important business. Thank you and goodbye!"

Then he wobbled home to watch his favorite TV show — *How To Make Bigger Cakes* — and sat in his comfy chair to write down the recipes.

6. Bigger Squeaks

A bitter wind was blowing hard up Biffletop Hill and Spinner was turning his arms to make electricity, but his squeaks were worse than ever.

Scratchy, screechy squeaks!

They were so loud, all the animals had to cover their ears.

They were so squeaky, everybody down in Biffletop could hear, except the Lord Mayor, who was too busy with his electric cake-mixer, trying out a new recipe for *Banana Cream Pie*, which he had written down while watching television.

In Café Dish-Dirt on Wottapong Street, Jack Throttle was sitting with three of his worst friends, but he liked them anyway.

"We don't need that turbine," Jack growled, putting a hairy hand over his hairy ear.

"Quite right," agreed Betty Baker, who was wearing red ear-muffs like big raspberries. "We get enough electricity from wooden poles with strong wires, so we do not need that little squeaky-turbine-squawk!"

Johnny Giblets looked up the hill like a rooster hunting juicy bugs.

"We must stop that noise! We've waited long enough for that lazy Mister Mayor. I have an idea…" he said, sly as a fox in a hen house, which is how he learned his tricks. He nudged the skinny fellow sitting alongside him and asked, "What do you reckon, Pete?"

Pete Potts worked at Biffletop Paint Factory, except when his boss wasn't looking; he had a mustache as thick as a brush and wore dungarees that did not fit, despite all the food he gobbled. "I reckon," muttered Pete, chewing a pig's trotter, and he gave a crafty smile, sneaky as a plotter.

Betty Baker closed her eyes tight and howled, "Did you hear that squeak? This is getting worse! I'm getting a headache."

"I'm getting an earache," added Johnny Giblets.

"Let's go," said Jack Throttle.

They scooted outside and climbed into his yellow taxi, which went zigzagging up the hill because Jack was hardly the best driver in Biffletop.

7. Smile, Please!

Spinner was turning his squeaky arms on Biffletop Hill when a yellow car stopped nearby and four people popped out like peas from a pod. They strode across the grass carrying all sorts, including a nice wicker basket.

"Strangers!" said Prickle Hedgehog, curling himself into a ball of sharp spikes just in case.

"They're here for a picnic," Spinner said. "And they're coming to say hello. That's nice."

But unfortunately, he was mistaken.

Johnny Giblets reached into the wicker basket and threw eggs that cracked against Spinner's head and smelled yukky. "Because they're two months old, take that!" said the foxy chicken farmer, throwing another.

Betty Baker hurled bags of flour that exploded in Spinner's face and made his eyes all dusty and his arms stick fast. Pete Potts used a paintbrush to write big red words on Spinner's tummy and, when it was all finished, Jack Throttle took a photograph and said, "Smile, please, Squeaky!"

How they laughed as they ran away. They climbed into the yellow taxi and drove off with even more zigzags, and you can probably guess why.

8. Red Paint and Rude Words

When the taxi had gone, the animals gathered around Spinner, with anxious looks. Tuft flew up and down, reading aloud the big words in red paint on Spinner's tummy, carefully announcing each letter, *"T-R-A..."*

"What does it say?" cried Spinner, whose eyes were too dusty to see.

"Trash This Turbine!" croaked Frank Frog.

"That's not very nice," added Flash Rabbit, with a sad face.

"But hardly surprisssing," hissed Snake Adam.

"Some picnic," said Big Badger, watching the taxi. It was just a little yellow dot in the distance now, swerving from side to side, like a busy bee.

Prickle Hedgehog sighed and added, "Those people did not come to say *hello*, they came to say *get lost*."

Spinner felt tears in his eyes and the flour made them sticky.

"Biffletoppers don't like me," he wailed. "I want to live by the beach, with Mom and Dad and my sister, Turner!"

"Don't cry," said Marjorie Mouse. "You just need to get cleaned up, and someone should put some oil in your tummy to stop those squeaks."

"But who?" asked Spinner, and even Tuft had no answer.

Purple clouds sailed over the valley, lightning flickered and thunder boomed. Flash hopped down a rabbit hole and yelled, "Storm coming!"

Spinner gasped, gazing fearfully at the darkening sky.

9. No Time for Munch-time!

Joe Driver finished work at midday and parked his truck outside the turbine factory. He washed his hands, opened his sandwich box and said,

"Munch-time, at last!"

He unfolded his newspaper — the *Valley Voice* — and stared at the photo on the front. It showed an unhappy little turbine covered in red paint, white flour and yellow eggs. "I know this face," Joe said. "This is Spinner on Biffletop Hill!" He read the headline, which had some big words in it:

PROTESTORS DEMAND REMOVAL OF TURBINE

Joe read it again slowly, to make sure he understood. "What's all this?" he said. "*Protestors?* That means angry people making a big fuss. *Demand* means when you ask strongly. *Removal* means when you move something away, and *turbine* obviously means Spinner. So, in other words, my newspaper says: *Angry people want Spinner moved away.* But why?"

Joe read on and soon lost his appetite because he was so concerned about his little friend on Biffletop Hill. "Spinner squeaks! How come?"

He folded his newspaper and scratched his head. He heard someone whistling and turned around. It was one of his workmates in the factory, squirting oil in the tummy of a new turbine: *squirt-squirt, whistle-whistle.*

"Now I remember!" Joe gasped. "It's my fault!" He closed his sandwich box. "No time for munch-time! I have to help Spinner!"

Joe ran to the storeroom where Riley Rules was checking supplies, assisted by a black cat that had caught every single mouse (except one).

"Hello, Riley," said Joe, out of breath. "I need a large can of turbine oil, please."

"You need to sign this form," replied Riley, offering his pen.

Joe quickly scribbled his name. "Done! And a stepladder, please?"

"First, sign the Stepladder Form," said Riley, "here and... here."

Joe scribbled his name. "Done! And a mop and bucket, please."

"Sign the Mop Form," insisted Riley, "and what color bucket?"

"Any color," said Joe. "Could you hurry up, please?"

"Of course, "Riley answered. "If you sign the Hurry-Up Form."

"Done," said Joe, signing here, there and everywhere.

Riley caressed his little black mustache and said, "Joe, is there a problem? Because if so, you'll need to sign the..."

Joe grabbed what he needed: a big can of oil, a stepladder, a mop and a bucket. "No problem, Riley, but if there is, I'll solve it. Bye!"

He raced out carrying his supplies, and jumped back into his truck.

He drove quickly along the winding roads to Biffletop and raindrops danced happily on his roof — *pitter-patter* — because a storm was coming.

"Don't worry, Spinner!" Joe said, "I'm coming to help!" He tooted his horn, hoping that a squeaky little turbine might hear it across the green hills.

10. Crack! Bang! Hiss!

On Biffletop Hill, the afternoon darkened into evening. The animals were all safely tucked up in warm homes and Spinner stood alone.

Cold rain came shooting down like bullets from angry-looking purple clouds.

The wind blew harder every minute, but Spinner could not turn his arms because they were sticky with flour and stinky with goop from rotten eggs.

"Why even try?" he mumbled. "I'm useless, just a broken turbine that nobody likes and everyone hates! Who cares if I never make electricity?"

He thought about Mom and Dad and sister Turner at a sunny beach, where seagulls flew and happy children played in the funfair. A big tear rolled down his mucky face.

Suddenly, lightning cracked from the grumbling sky and struck a tree, which exploded in flames. Spinner watched the hungry yellow fire licking at blackened branches and the thunder roared across the valley: *GOOD JOB!*

Soon, another terrible flash lit the hillside, as if a star had crashed from the sky. Spinner closed his eyes and the thunder boomed: *GOOD JOB!*

When he finally dared to peep, Spinner saw the strangest thing: one of the strong wooden poles that sent electricity in wires to Biffletop had blown over and now lay flat on the ground, buzzing and spitting hot white sparks.

The sparks fizzed along the black wires to the next pole, which cracked and banged and hissed as it wobbled and fell; then the next pole, and the next, one after another: *Crack! Bang! Hiss!* The wooden poles looked very worried as their wires stretched and snapped and twisted in the wet grass like fiery worms.

Spinner peered through the heavy rain and noticed something else, too: the bright lights of Biffletop were going out one by one, and that busy little place was getting darker every second. He quickly realized why: if the wires were snapping up here, electricity could not reach the town down there!

BAD JOB!

11. Darkness Falls

*T*he Lord Mayor was at home dozing in his comfy chair and dreaming of Extra Large Apple Pie, when he was awoken by a strange noise: *Crack! Bang! Hiss!*

"Oops, I fell asleep," he said. "Because I work too hard. Oh well, time to polish my nice gold chain."

He stood up and looked in the mirror, but it was too dark to see.

"Who turned out my lights?" he asked, trying the switch. *No good.* He checked the plugs. *No luck.* "Is this a blackout?" the Lord Mayor wondered.

He sat down and munched cake and thought about what a blackout would mean. *No more recipes on TV?* In fact, the more he thought about it, the more worried he got, because a blackout would mean no electricity for:

Cake-mixers
And microwaves
And computers
And mobile phones
And the hospital
And the cinema
And the Old Folks' Home
And traffic lights
And Biffletop Airport
And recipes on TV *(oops, already said that)*

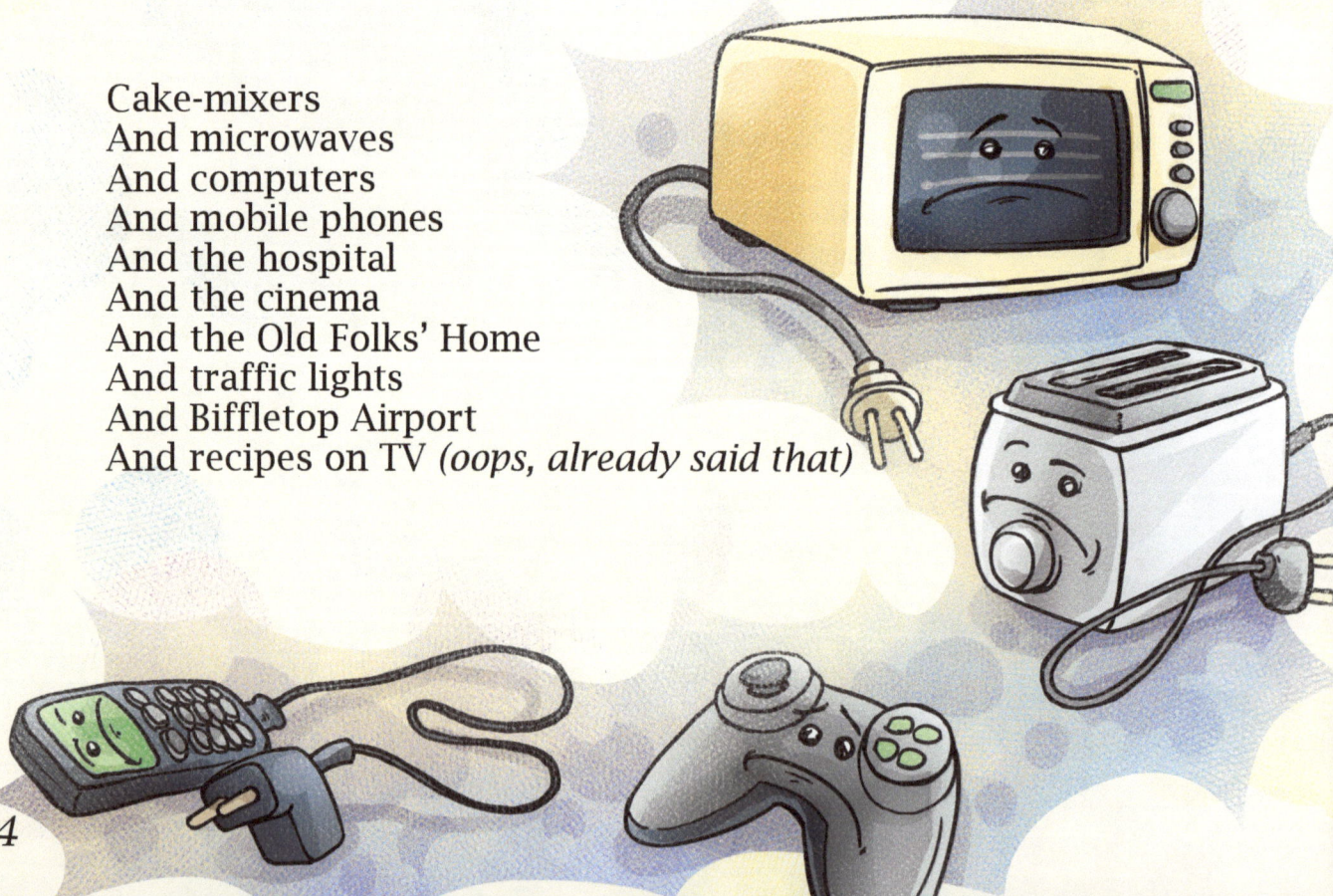

The Lord Mayor walked about bumping into things and talking anxiously to himself. "What will folks do? What will they say?"

He was about to find out. He peeped through his window and saw lots of unhappy people in the street, holding fiery torches. Some were even waving their fists.

"More complaints!" he whimpered. "I'll go and talk. I do hope someone has a plan for such an emergency!"

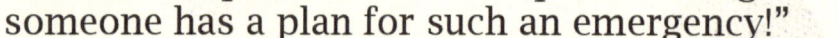

12. People Power

"Hello everyone, nice to see you!" said the Lord Mayor, from his doorstep.

"You wouldn't be able to see us, if we didn't have these torches," chuckled Charlie the Chief Engineer, "it's a total blackout."

"A total mess," groaned Norbert Spratt from the fish shop. His oily hair was the color of red snapper and his crabby kids slept five to a bed, like sardines in a tin.

"What are we going to do?" asked the Lord Mayor.

The people huddled closer. Some of them had scared faces, some had angry faces, some had faces you could hardly see, it was so dark. In fact, it was so dark, the ghosts in Biffletop Castle were scared out of their flapping bed sheets, and their spooky doctor advised them to lay off the boos.

"Lord Mayor," barked Miss Cane the headmistress, "or, perhaps I should say *Lord Moron*. Anyway, what are *you* going to do? Speak up; don't snivel like a snipe! If that squeaky turbine were working, we wouldn't be in the dark; but it isn't, so we are. Why didn't you fix it, as you promised?"

The Mayor shrugged. "Because I was busy working and… stuff."

Charlie the Chief Engineer asked, "Did you find out who threw moldy flour and stinky eggs? It jammed the turbine. They made things worse!"

The Lord Mayor shook his head, wondering how to make things better. His little legs went all wobbly so he sat down on his doorstep and Charlie said, "Do you have a plan for this emergency, Lord Mayor?"

Everyone waited, hoping for a good answer. The Lord Mayor looked up and said, "No, but I have a recipe for Banana Cream Pie, which I'll share tomorrow, if you could stop asking questions and go away. How's that?"

The Biffletoppers agreed to go away. They also agreed to throw the Lord Mayor in Lake Biffle (which, as you know, is quite wet).

"But I can't swim!" he yelped, as they carried him on their shoulders.

"Don't worry," said Norbert Spratt, "you'll float."

13. Yum-Yum, Oil in My Tum!

Spinner peered through wind and rain on Biffletop Hill to see headlights twinkling towards him, and mumbled sadly, "More unfriendly visitors?"

He saw a red truck stopping. Joe Driver jumped out and came running across the slippy-uppy grass with a stepladder, a mop, a bucket, and a big can of oil. "Joe!" Spinner yelled happily, "you came back!"

"To clean you up and get you started!" shouted Joe. The purple clouds circled with mean looks as Joe washed off stinky egg, sticky flour and red paint. He squirted turbine oil in Spinner's tummy and said, "Sorry I forgot!"

Spinner giggled as the oil tickled him inside. It was soon flowing through his pipes and sprockets and dribbling on his cogs and crankshafts, until at last his arms began to turn, slowly at first, with no squeaks, in the fierce wind of the biggest storm ever to batter Biffletop.

"You'll have to work extra hard, there's a blackout," said Joe, "Biffletop has no electricity because the wooden poles fell down!"

"Can't you fix them?" asked Spinner.

"Not until the storm ends!" replied Joe. "How's that oil, are you feeling better?"

"Much better and less squeaky!" replied Spinner.
"Great! Let's see what you can do!" Joe said.
"I'll do my best!" shouted Spinner. "Thanks, Joe!"

The angry purple clouds raged and roared, doing their worst, trying to topple the brave little turbine. But the harder the wind blew, the faster Spinner's arms spun, and soon they were just a silver blur in the black night.

Joe climbed back in his cab and sipped hot coffee as he watched; it was quite a battle out there and he said to himself, "I wonder who will win?"

14. A Lucky Escape

When the crowd of angry Biffletoppers arrived at Lake Biffle, someone yanked the gold chain from the Mayor's neck and said, "You won't be needing this, Lord Lazybones!"

They were just about to throw the Mayor in the water (which as you know is quite deep), when something unexpected happened: the lights began blinking, on and off, all over the town, because the electricity was coming back!

First the traffic lights: *blink-blink!* Then the lights outside the cinema: *blink-blink!* Next, the lights in houses: *blink-blink!*

The Biffletoppers dropped the Lord Mayor on his bottom and ran home cheering and yelling, "The lights are on! We're saved! No more blackout!"

Charlie the Chief Engineer rubbed his chin and wondered *how come?*

Meanwhile, up in the hills, Sergeant Slammer of Biffletop Police was standing on a dark road next to a yellow taxi. It had crashed into a tree and its driver, Jack Throttle, was fast asleep with a nasty bruise on his head.

Sergeant Slammer knew a thing or two about this Throttle fellow. For example, Jack would have a headache when he woke up, and a surprise, because a tiny mouse had sheltered in his hairy ear during the storm and seemed quite happy to stay there.

Sergeant Slammer also noticed a stinky cracked egg and some moldy flour in the taxi, which soon got him thinking, so he scribbled in his notebook: *Wot I fink is…*

15. Spinner the Winner!

Next morning was bright and sunny. All the birds were singing happily and even a worm came out for a wiggle and didn't get pecked.

Spinner was fast asleep after whizzing all night, but soon awoke when he heard the happy voices of people from Biffletop coming up the hill.

"People want to say sssorry," hissed Snake Adam, and all the animals waited to see if something special would happen again today, and of course, it did.

The Biffletoppers gathered around the turbine and Charlie the Chief Engineer said, "Little Spinner ended the blackout, thanks to Joe Driver's quick thinking!"

Everyone clapped and Miss Cane suggested, "We need a new Lord Mayor. I think it should be Joe. Do we all agree?"

The Biffletoppers shouted *YES!*

They put the shiny gold chain around Joe's neck and he posed proudly for a photo. "Thank you!" he said. "But don't forget who really saved Biffletop! In a fierce battle with that nasty storm, Spinner was the winner!"

The Biffletoppers cheered and all agreed that Spinner was a strong little turbine who did not spoil their view, and who did not have squeaky arms any more, *no way José.*

51

José worked in the Oil Department of the turbine factory and he promised Spinner three barrels of *Super-Kreemi*. "You deserve it!" he said.

"And a holiday too, I think!" added Lord Mayor Joe. "So, where would you like to go, Spinner?"

"Guess!" said the little turbine, with a big smile for the TV cameras.

Meanwhile, in Biffletop Jail, Betty, Johnny, Pete and Jack confessed to their crimes and offered to apologize to Spinner the Winner, just as soon as Sergeant Slammer let them out (if he ever did).

There was one last thing for Spinner to do. "Goodbye and thanks for being my friends," he said to the animals, "I'll never forget you."

"Bessst of luck," hissed Snake Adam, with a slippery smile.

"I hope it's windy, wherever you're going," sighed Prickle Hedgehog.

"And thanks, you did a good job!" added Marjorie Mouse.

"Time to hoot!" said Tuft Owl, flying away. "Bye, Spinner!"

16. Back to the Beach

Soon Spinner was back in Joe's truck, riding towards the ocean. He saw a seagull flying high on the wind, blue water twinkling in the distance and lots of big white turbines standing on high cliffs above a golden beach.

He watched their long arms circle gracefully in the breeze and he shouted, "Mom! Dad! I'm back!"

"Hey, Spinner!" yelled his sister, Turner.

Joe parked the truck and hopped out, wearing a shiny gold chain, and Riley Rules could hardly believe his beady eyes.

"I hear this little turbine is quite useful," Riley said, looking in a big black book. "And that's good news, because I'm expecting lots of people at the funfair, so we'll need lots of electricity. Maybe this little one should stay here, if the Lord Mayor agrees."

"Who, me?" asked Joe, brushing a speck of sand from his gold chain.

"Yes, Lord Mayor."

"I agree," replied Joe.

"Yippee!" whooped Spinner, twirling his arms as the big yellow crane lifted him into a nice soft hole in green grass. "But what about Biffletop?"

Riley gazed far across the hills and said, "I'll send some turbines."

"Good idea!" replied Spinner, and a seagull that was standing on Riley's hat flew away to tell Tuft the Barn Owl.

"Thank you, Joe! You're a nice Lord Mayor!" said Spinner.

"And you're a strong little turbine!" replied Joe. "Bye for now!"

Joe drove away and waved his cap, which was not green and spotty today. It was made from beautiful red velvet and you probably know why.

Spinner gazed at the blue ocean and a cool salty wind came across the white-tipped waves to blow any bad memories away. The sun looked like a big orange balloon floating in the sky, and soon Spinner's tummy filled with a nice feeling that was not turbine oil or electricity. It was happiness!

Spinner was happy at last, because he could hear children laughing and whooping as they rode in bumper cars and scary rides at the funfair.

But most of all, he was happy to be with his family, and, as you may know, that feels nice.

"Welcome home, little Spinner!" sighed Mom, with a tear in her eye.

"Good to have you back!" shouted Turner, "we've got work to do!"

"And how do you like the view?" asked Dad.

"I love it!" said Spinner, whizzing his arms around, "I love it!"

*Feeling artistic? Ready to have some fun?
If so, color in the outlines below!*
Tip: for inspiration, find the original page in Spinner the Winner.

Joe is driving home, after a job well done.
To make his trip colorful,
grab your tools and get to work!

Want to work on more pages? Look for
Spinner the Winner - Coloring Book.
It's available now!

www.ingramcontent.com/pod-product-compliance
Lightning Source LLC
Chambersburg PA
CBHW051028180526
45172CB00002B/501